A Silent Tsunami

The Urgent Need for Clean Water and Sanitation

This report is available on line at:
>	www.aspeninstitute.org/EEE/water
>	and
>	www.env.duke.edu/institute/water

For additional copies of this report, please contact:

>	The Aspen Institute
>	Publications Office
>	109 Houghton Lab Lane
>	P.O. Box 222
>	Queenstown, MD 21658
>	Phone: (410) 820-5326
>	Fax: (410) 827-9174
>	E-mail: publications@aspeninstitute.org

For all other inquiries, please contact:

>	The Aspen Institute
>	Program on Energy, the Environment, and the Economy
>	One Dupont Circle, NW
>	Suite 700
>	Washington, DC 20036-1193
>	Phone: (202) 736-5857
>	Fax: (202) 467-0790
>	E-mail: katrin.thomas@aspeninst.org
>	Web: www.aspeninstitute.org/EEE

Copyright © 2005 by The Aspen Institute

The Aspen Institute
One Dupont Circle, NW
Suite 700
Washington, DC 20036-1193

Published in the United States of America in 2005
by The Aspen Institute

All rights reserved

Printed in the United States of America

05-015
ISBN: 0-89843-435-1

Table of Contents

Foreword v

Recommendations in Brief 1

"A Silent Tsunami" 3
   Report of the Co-Chairs

Letter to Organizers, Fourth World Water Forum 31

Participants 37

Additional Information 40

Foreword

Few issues matter more to public health, economic opportunity, and environmental integrity than the availability of clean water and sanitation. With the 4th World Water Forum scheduled for Mexico City in March 2006, the Aspen Institute and the Nicholas Institute for Environmental Policy Solutions at Duke University conducted a multi-stakeholder dialogue to help highlight the importance of global water issues, suggest steps to provide services more rapidly and effectively, and to identify and draw attention to constructive ways the US government and other US participants can take part in the Forum.

In April 2005 a distinguished group of leaders from business, government, and environmental and other non-governmental organizations met for two and a half days at the Aspen Institute's Wye River Conference Center to learn from each other, to explore the sometimes competing values underlying policy disagreements, and to consider appropriate responses to the challenges identified. Their expertise was matched by their commitment to discover and implement solutions to the world's water challenge. This report is a summary of their conclusions.

We were honored to have as our distinguished co-chairs Ambassador Harriet C. Babbitt, Senior Vice President of Hunt Alternatives Fund and former Deputy Administrator of the Agency for International Development, and William K. Reilly, Founding Partner of Aqua International Partners and former Administrator of the Environmental Protection Agency. Their varied and complementary experiences – in development and the environment, in the public and private sectors, in Administrations of both political parties – and their ability to explore details while framing the issues broadly allowed them to extract and focus the wisdom of an extremely knowledgeable group.

The co-chairs' overview and conclusions constitute the body of this report. While the richness of the dialogue cannot be recreated, their essay summarizes their sense of the group's principal conclusions and insights. The findings and recommendations are based on the group's agreement in the concluding session of the meeting, but participants were not asked to agree with their final wording, and no person's participation should be assumed to imply his or her organization's endorsement of any specific finding or recommendation.

We appreciate the encouragement of the Mexican organizers of the 2006 World Water Forum in the conception and organizing of the meeting and their receptivity to our findings. We gratefully acknowledge the assistance of Mark Van Putten of *ConservationStrategy*™ and Gordon Binder of Aqua International Partners in planning and preparing for the meeting, and the grace and efficiency with which Katrin Thomas handled advance arrangements and managed the details of the meeting.

We are also grateful to the Charles Stewart Mott Foundation, the Wallace Genetic Foundation, Procter & Gamble, Coca Cola, American Water, Dow Chemical, and CH2M-Hill for their financial assistance. Without their generosity and confidence in our work, the meeting could not have taken place.

The mission of the Aspen Institute is to foster enlightened leadership and open-minded dialogue. Through seminars, policy programs, conferences and leadership development initiatives, the Institute and its international partners seek to promote nonpartisan inquiry and an appreciation for timeless values. Its Program on Energy, the Environment and the Economy brings together in a neutral forum individuals from government, industry, environmental and other public interest groups, and the academic world to improve policy making through dialogue on current and future energy and environmental issues.

The Nicholas Institute is a new and innovative environmental science and policy institute designed to fill the need of policymakers, businesses, and organizations for unbiased data and dialogue about the important environmental questions facing the world. The Institute will marshal the broad resources of the Duke University community – including the Nicholas School of the Environment and Earth Sciences, the nationally ranked Fuqua School of Business, and the Duke Law School – and the expertise of university partners in industry, government, and environmental organizations to craft innovative and practical solutions to critical and inevitable environmental challenges.

John A. Riggs	Timothy Profeta
Executive Director	Director
Program on Energy, the Environment, and the Economy	The Nicholas Institute for Environmental Policy Solutions
The Aspen Institute	Duke University

Recommendations in Brief

1) Clean water and sanitation must become a higher priority because they are fundamental to human health and reducing poverty.

2) All schools and orphanages should have clean water, sanitation, and hygiene education by 2015.

3) The President of the United States and his Administration should develop a strategy to mobilize American resources and institutions to become more involved in water internationally.

4) For reasons of health, the economy, and environmental sustainability, governments must invest more in water infrastructure.

5) Decisions about covering the costs of clean water and sanitation should be decided through a participatory process that ensures the needs of the poor are met and provides sufficient funds for maintenance.

6) Because water and sanitation are often the responsibility of women in much of the developing world, they should become more directly involved in managing water resources and making water-related decisions.

7) Development assistance should emphasize building local capacity, creating legal frameworks for managing water, and building local sources of funding.

8) Promising partnerships among governments, not-for-profits, community and faith-based organizations, and businesses should be replicated and scaled up.

9) Decentralized water treatment systems or point-of-use household treatment, coupled with sustained hygiene education, should be deployed more widely, especially where they can reduce water-related disease immediately.

10) Decisions about managing water resources must involve all stake holders and all relevant factors in supply and demand, with efficient water use and protection of ecosystems as central goals.

A Silent Tsunami: The Urgent Need for Clean Water and Sanitation

William K. Reilly and Harriet C. Babbitt, Dialogue Co-Chairs

Across much of the developing world, a silent tsunami is raging: for lack of clean water and sanitation, as many poor people are dying each month as perished during the Southeast Asian tsunami of December 2004. An estimated 6 million died in 2003, according to the World Health Organization, many of them young children. In addition to death and illness, a loss of hope and opportunity are direct consequences of water-borne and related diseases. But unlike the tsunami that devastated Southeast Asia, this one can be stopped.

Access to adequate, clean, affordable supplies of water, as well as sanitation and hygiene, is fundamental to human health, to human dignity, to reducing poverty, and to expanding economic opportunity. Yet a billion people or more go without safe drinking water; twice that lack adequate sanitation.

In the past, the conventional response might well have been to plan large engineered drinking water and wastewater facilities, to lay pipes and extend coverage to each household. That is a lengthy, expensive, and difficult proposition. And there are many impediments, from insufficient project development capabilities to financial risks to local opposition, that explain the lack of viable projects. Such large projects may still make sense in densely populated areas, but new approaches are necessary to get clean water and sanitation to people in villages and other non-urban places. Both water and sanitation are critically important, although each represents a different challenge for service providers and a different calculation of costs and benefits.

Fully satisfying the need for clean water and sanitation on a lasting basis requires a perspective broader than just delivery of basic services. In too many countries, water resources overall are badly managed. Responsible ministries are weak or lack capacity. Local water utilities also lack adequate skills and resources but nevertheless see the responsibility for water and sanitation devolve to them. Investment in water infrastructure is limited. Since water tariffs are minimal, there is insufficient revenue even to maintain the system. And the threat of climate change complicates the challenge. It has the potential to upend familiar patterns of precipitation, leading to drought or more flooding and rendering existing infrastructure obsolete.

In short, providing clean water and sanitation and sustaining the economy and the environment require better management of water resources at all levels of government. To meet the challenge internationally, it is difficult to escape the conclusion that more national governments are going to have to elevate the priority for water in their budgets, development plans and projects, and other decisions. International donors, public and private, will also have to step up their efforts. And yet, in contrast to the outpouring of support in the Southeast Asian tsunami's aftermath, the political will and other essential elements to address water needs seem, for the most part, in short supply.

The ultimate responsibility for providing safe, affordable, and ecologically sustainable water and sanitation services falls to governments, at the national, provincial, or local level. How these services are provided - whether through public utilities or private operators, through concessions or community groups - matters less than that the services are being delivered.

Donors, bilateral and multilateral aid agencies, and others have a critical role in meeting the water challenge, especially in helping poor countries and in fostering regional cooperation on water issues. Money will be needed, but that is not the only contribution donors can make. They can also help fill the need for technical help in creating legal and regulatory frameworks and long-lasting institutions to improve water management; for technology transfer; and for exchanges, education, and training to build capacity. These may offer low-cost means of providing assistance, especially for development agencies whose budgets are spread thinly to meet many legitimate purposes.

Still, we heard repeatedly in our dialogue and we have come to accept that in all but the poorest countries most of the money spent on water inevitably will have to come from within the affected countries themselves. That means finding innovative ways to mobilize and put to work local or domestic financial resources. And it means enlisting nongovernmental, community, and faith-based groups, as well as the business sector, in creative partnerships to deliver needed water services. In the end, we concluded that meeting this challenge can wait no longer.

Solutions Are Available

The group of experts who gathered at Wye are intimately familiar with the sobering array of issues and statistics, and we chose not to belabor the magnitude or complexity of water problems. For us, the most illuminating part of the discussions was learning about the rich examples of projects and sponsors bringing safe, affordable, and sustainable water and sanitation to those in need. A good amount of experimentation is under way with approaches that go beyond delivering water through large-scale, expensive engineered projects. Several creative models, institutional reforms, innovative financing, and partnerships between and among development assistance agencies, nongovernmental groups, and private companies were described. Especially intriguing are decentralized water treatment systems and household point-of-use products that offer immediate intervention to reduce death and disease as well as community and faith-based models for extending access to services. (See Box, Promising Examples and Models, pp. 8-14)

Many of these projects are promising. To reach more people, however, they need to be expanded, replicated, and scaled up, no easy task to be sure. They will need money and technical know-how, which may become increasingly available through governmental support - in the United States, through the US Agency for International Development (USAID), USAID's Global Development Alliance (GDA), and the new Millennium Challenge Corporation. And around the world, the private sector, non-governmental groups, philanthropies, UN agencies, multilateral development banks, and others have significant contributions to make, financial and otherwise.

The results of these projects are for the most part going unheralded. We heard again and again of the need for a simple, compelling message that could draw attention, raise awareness, build public support, and, most importantly, mobilize resources and motivate action. We learned about the need to tell the stories of people, families, children who have benefited from better access to clean water and sanitation. We heard, too, about the need for variations on the theme that could engage new audiences across different sectors of society. Several participants in the Wye session tested messages about saving lives and expanding opportunities. But we quickly came to the realization that this gathering of technical experts and policy advocates was probably not the best group to devise messages to spur people to act.

One singular thrust, however, caught everyone's imagination: We were captured by the potential impact of using schools and orphanages to mobilize resources to deliver clean water and sanitation to children. UNICEF, we learned, recently estimated that half the world's schools lack these basic services. The task seems manageable, something that could be pulled off within a reasonable time frame, even if not everyone in need would be reached right off. Aside from the obvious health benefits for children, this is seen by community and faith-based groups with direct experience as a way to improve school attendance and academic performance, to give children reason to hope for a better future, and a means to benefit their families and their communities through outreach and expanded access to water. School attendance, especially by young girls, who now may spend hours each day hauling water from distant sites, would likely increase with all the collateral benefits this would bring to societies. This is a real opportunity to engage more fully government, business, civil society, and others to assist those poor countries where the political will to address water issues is beginning to emerge.

The humanitarian impulse to get clean water and sanitation to people in need is strong. The economic and environmental arguments compelling. The array of solutions is growing. The time for action is now.

PROMISING EXAMPLES AND MODELS

- Mexico recently enacted a new national water law under which responsibility for water decisions is decentralized and intended to be made involving local officials on a river basin or watershed basis. Mexico for the most part has already reassigned responsibility for irrigation from the national government to the irrigators themselves, a move that can improve water management and bolster civil society. No one underestimates how difficult change is in this country where management of water resources has been highly centralized. But changes are occurring and considered by those who know the country nothing short of revolutionary.

- Globally a growing number of nongovernmental groups are employing community-based models to provide water and sanitation. This approach involves working closely with affected communities, tailoring projects to local water conditions, tapping indigenous knowledge, applying inexpensive and convenient technologies, ensuring that services are sustainable financially and operationally, and integrating hygiene education into all projects, especially the health value of regular hand washing. Water for People, for example, is the not-for-profit arm of the American Water Works Association, which represents water utilities. A little more than a decade old, with a modest, but growing budget, Water for People has been working in 450 communities on every continent.

Water for People has learned in its work in Africa that often the first step is to build trust between the people of the community and the institutions of government; then comes the technical and other assistance that actually begins to deliver clean water. Sanitation, the field staff have learned, is a bigger challenge. The organization's experience underscores several other critical points: gender issues are paramount throughout the developing world, because water and sanitation are often the responsibility of women and girls. They've also learned to insist that the community contribute something of itself, in labor and importantly finance - or the endeavor isn't valued. Partnerships are critical and can tap Peace Corps and other volunteers to bring in needed skills. Water for People is organizing a "sister city" exchange whereby US water utilities will help build technical capacity in local water providers.

- Living Water International is one of a number of faith-based organizations involved with providing safe drinking water. Drawing on its pool of volunteer engineers, geologists, construction managers, educators, and others, and working especially with schools, orphanages, and hospitals, Living Water drills wells, provides pumps, trains local people in maintenance and repairs, and offers related services such as hygiene education and mobile medical units. To date the group has completed more than 1800 water projects, serving over 3.5 million people daily in 21 countries.

  Living Water International also was a prime mover in creating the Millennium Water Alliance to coordinate the clean water and sanitation work of several community and faith-based providers, with a goal of reaching 500 million people by 2015.

- World Vision, which has helped provide schools with water in Cambodia and West Africa for more than 15 years, is part of the West Africa Water Initiative, launched at the World Summit on Sustainable Development in 2002. In Ghana, Niger, Mali, and elsewhere (with support from the US Agency for International Development, the Hilton Foundation, and others), the Initiative will bring water to 400,000 people by identifying and developing water supplies, building local water management capabilities (including repair and maintenance), and devising a self-sustaining means of finance. With many partners, it's taken a while to plan, but activities are now under way.

- Winrock International is one of the partners in the West Africa Water Initiative. In Nepal, Winrock International also carried out a project to install drip irrigation that benefited not only farmers, but water providers, manufacturers, and the broader community.

- RWE Thames Water, which primarily provides water services in England, also holds water concessions in several cities around the world. Learning from the enormous expense of extending coverage in Jakarta, Thames began to explore lower-cost ways to deliver clean water. As one example of its subsequent approaches, Thames recently joined with the non-profit community-based organization WaterAid, CARE, and other non-profit, for-profit, and academic organizations in a partnership called Water and Sanitation for the Urban Poor (WSUP). The focus is on delivering sustainable, equitable, and affordable water and sanitation services in low-income urban and peri-urban communities, among the most challenging environments in which to operate. The first project is getting started in Bangalore, India, with sponsors expecting to announce soon the next effort in Africa.

These efforts are targeting areas of greatest need, building local capacity by involving the community from the start, using donor funds to jump-start the project while designing self-supporting and sustainable operations, and incorporating hygiene education and integrated water management as key elements.

- Widely known for its humanitarian work in disaster relief and development, CARE has worked in the water sector for nearly half a century, helping reach over 20 million people in more than 40 countries. CARE, too, employs a community-based model and incorporates water management principles at the local level in places like Bangladesh, El Salvador, Jordan, and the West Bank, to cite a few.

  With the US Centers for Disease Control and others, CARE helped pioneer the household treatment Safe Water System, combining a simple low-cost disinfectant, safe storage vessels, and hygiene education. During the tsunami in Southeast Asia, CARE brought immediate help with point-of-use treatment to get people clean water. As tragic as the impact of the disaster was, millions lacked access to clean water and sanitation before the tsunami. After attending to immediate and urgent needs, CARE and other groups will use the recovery and rebuilding phase to develop community water and sanitation services in the region on a lasting, financially viable basis.

- CARE also put to good use a product created by Procter & Gamble in partnership with USAID, Population Services International, and others. PUR is a sachet of simple chemicals (the equivalent of a waste water treatment plant in a small packet) that when mixed with dirty water destroys or settles out pathogens, heavy metals, and other contaminants, providing clean drinking water at the household level.

The company's intent at the start was to develop a commercial product that millions of poor households could readily afford, called by marketing experts "the bottom of the pyramid," but market introduction costs were too high for a conventional business model to succeed. The company then created partnerships with several governments, social marketing NGOs, and global relief organizations to make the product available for disaster relief, including the 2004 tsunami, and to create small scale local entrepreneurial business models that benefit rural villages and urban slums in the developing world.

- The Coca-Cola Company is also taking a keen interest in water issues in its operations and in communities near its facilities, surveying more than 850 facilities in over 200 countries to identify priorities related to watershed stress, supply reliability, and stakeholder issues.

Coca-Cola also has started a number of water projects, working with local communities and other partners. In 2004, in Vietnam, the company and the United Nations Development Program launched Clean Water for Communities to develop sustainable solutions that meet community needs. The project provided 180 water tanks to nearly 500 families in six provinces, giving them access to clean water. In India, Coca-Cola has begun harvesting rainwater in all company plants and with government authorities set up local rainwater harvesting projects around the country. In Rajasthan, for instance, the company joined with a local NGO to set up the state's first such project. In Kadadera, an indigenous system of water collection was rehabilitated through the construction of more than 30 recharge shafts. Elsewhere, as part of Coca Cola's assistance to tsunami relief, the company earmarked $1 million for sustainable development of water and sanitation in affected communities through well rehabilitation, village water system installation, and water storage. Resources were doubled through a partnership with the UN Foundation and UN agencies.

Finally, Coca Cola is working with World Wildlife Fund and local partners to fund watershed conservation projects in the Chihuahuan Desert straddling the Texas-Mexico border; the Mekong River Basin of Vietnam; the Atlantic forest of Brazil, the Zambezi Basin in southern Africa, and the rivers and streams of the southeastern United States.

- WaterHealth International (WHI), a new US company, is developing two different commercial models for rural and urban communities. Decentralized community water systems are providing clean water in quantities that can serve rural populations of up to 3,000 people. In its first installation in the state of Andhra Pradesh, India, in a partnership with the Naandi Foundation, treated water is now reaching 80 percent of the households in the village, and residents for the first time are paying small fees. The company estimates that the fees should be sufficient not only for operating and maintaining the facility, but also for recovering the initial capital investment over a period of years. WHI's franchised water stores in the Philippines provide opportunities for local entrepreneurs to own and operate small businesses that produce and deliver clean water to residents in urban and peri-urban communities. The company plans to expand both models to Africa and other regions.

- Although funding is concentrated mostly in the Middle East - Egypt, Jordan, and the West Bank - we heard about many projects that USAID has seeded directly and through the Global Development Alliance. All told, USAID's Water for the Poor Initiative has funneled money to more than 70 countries, helping more than 10 million people gain improved access to clean water and sanitation.

USAID has pioneered the use of loan guarantees in Tamil Nadu, India, to reduce risks for private lenders to water and sanitation projects. In India, too, and Mexico, USAID is helping develop pooled funds to support water projects, an idea borrowing heavily from the successful use of state revolving funds in the United States. USAID also has launched the Balkans Infrastructure Development Facility (BIDFacility) to support development of infrastructure, including water projects, in the region. Not every country may be ready now to take advantage of these innovative financing regimes. Yet experience in the United States suggests that starting at a small scale, with incremental pilot projects, can lead to bigger endeavors as experience builds.

Why Now?

Why indeed? So much has been written, so much has been said about water. The issue appears in news stories almost daily and has been catalogued in books and weighty reports. Awareness is rising about the urgency of the problems and the need for action. What's less well publicized, in our view, is that there are fixes to some vexing problems in some places - a chance, in other words, to stop the relentless tsunami.

We do not dismiss the good work in prior efforts. An international decade for drinking water supply and sanitation expanded coverage during the 1980s. Progress, however, was overtaken by population growth and urbanization as more poor people concentrated on the edge of cities with no access to public services. Another water decade was launched March 22, 2005 - "International Decade for Action - Water for Life" - to stress the central role that water plays in sustaining human life and well-being, to refocus political and global commitments on water, and to further cooperation at all levels. Three World Water Forums also have shed light on the global challenge in providing clean water and sanitation. A fourth is scheduled in March 2006, in Mexico City, the first time the international gathering will be hosted in the Western Hemisphere.

Further impetus has come from the Millennium Declaration and the Johannesburg Plan of Implementation, adopted by the community of nations during the 2002 World Summit on Sustainable Development. The Millennium Development Goals seek to reduce by half by 2015 the number of people without access to improved drinking water and adequate sanitation. Each in itself is a challenge, together something formidable.

Our colleagues at Wye reported that the drinking water goal is barely on target and sanitation is falling behind. This is cause for concern. UN agencies are doing a better job tracking implementation of the goals, but progress varies greatly region by region, with most anxiety expressed about the countries in Sub-Saharan Africa. China and India, too, are singled out for attention, as half the people without access to clean water and sanitation reside in these two most populous countries. The state of play certainly will be a focus at the upcoming World Water Forum.

In the United States, one especially promising development is the introduction of legislation on global water issues by Senate Majority Leader William Frist, co-sponsored with Senate Minority Leader Harry Reid and Senator Richard Lugar, chair of the Foreign Relations Committee. This makes for a propitious moment: For the first time there is leadership in Congress to spur greater US involvement in water issues internationally.

Dr. Frist encountered the water issue, we were told, during a trip to Africa to investigate HIV/AIDS, one of the other great tragedies stalking the African continent. He grasped readily the tie between HIV medications and the need for clean water if the disease is to be treated effectively. His legislation calls for, in part, a strategy to deepen US government support and to broaden the government's reach through partnerships with non-profit and nongovernmental groups, with companies, with the philanthropic community, and other sectors of US society.

The rationale for a greater US role is strong. For some who gathered at Wye, the humanitarian impulse is the driver - saving lives, reducing disease, giving people a new chance at life. For some, national security looms large. They see access to clean water and sanitation as building blocks for economic growth, political stability, and democracy, thereby reducing the breeding grounds for would-be terrorists and lessening the likelihood that unrest will draw the United States into far-off conflicts. For others, it is economic arguments - providing a climate for investment and job creation, tapping new markets for products and equipment, boosting productivity through better worker health, mitigating operational risks for company facilities that depend on water where it is becoming scarcer.

Still others at the meeting want to safeguard natural resources, because healthy, functioning freshwater and related ecosystems, including wetlands, estuaries, forests, and others, are directly linked to the well-being and future prospects of people.

Whatever the motivations, attention to the global water challenge today is growing. The consequences of lack of access to clean water and sanitation throughout the world are preventable. Given the urgency and the building momentum, the time for action is now.

The Broader Context

Demand for clean water is exploding everywhere. Population doubled over the past century, while water use grew sixfold! More people are concentrating in urban areas where there are few public services and little money to provide them. Experts forecast growing demand for water to meet food needs, even though agricultural production already takes the lion's share of water in virtually every developing country. Energy demand is growing rapidly, which puts a larger call on water resources. And economic development to produce everything from cars to microchips also requires clean water.

Rising demand is coupled with greater scarcity and other supply problems in many places. Water supplies are limited by pollution, excessive pumping of groundwater, mismanagement, outright waste, and inefficient use. Typically, only a small percentage of wastewater is treated before disposal - an estimated 14 percent in Latin America.

For years, development economists favored roads, ports, airports, power plants, and telecommunications as the building blocks of a growing economy. Water was hardly a priority. That began to change in the 1990s, as cholera swept through Latin America, reaching as far north as the US-Mexico border. More than 10,000 people died and a million were sickened. That got attention. Since then, new analyses of costs and benefits are helping to make the economic case for investment in water in developing countries.

The water challenge is not strictly a problem of the developing world. The United States is an affluent country that has benefited from decades of investment in water infrastructure. And yet, state water managers foresee scarcity even with normal precipitation. Analyses by several groups over the past few years project substantial funding gaps in meeting water infrastructure needs - hundreds of billions of dollars over the next 20 years for repairs of aging systems, upgrades and extensions, and meeting new regulations to deal with pathogens and other contaminants in the water.

The United States has learned a lot about water, in many instances the hard way when something didn't work or officials failed to consider the full array of factors, especially environmental factors. Healthy, functioning ecological systems are necessary for human and economic health. The price of failing to heed this is steep: rivers no longer reaching the sea, 50 percent of wetlands lost and with them water filtration and the buffers they provided against flooding, 20 percent of freshwater fish endangered, deterioration of coastal resources, and more. Costly fixes are required to compensate for the ecological services nature once provided. Today, expensive large scale, multi-stakeholder restoration initiatives, supported by federal and state funds, are under way for the Everglades, the Great Lakes, Chesapeake Bay, and other parts of the country.

But there's a lot that has gone right, and the point to stress is that investments by the US federal government in water infrastructure, and the institutions that plan and carry out these investments, have proved essential to the country's development. Of particular interest at our forum was a discussion of the Tennessee Valley Authority (TVA), which targeted a region of the United States lagging behind in development. In the 1930s, more than 90 percent of the people had no electricity, about three-fourths no piped water, few had radios, less than a quarter owned cars or trucks. Most lived on subsistence farming. Soil erosion and flooding were ruinous. Then came TVA. Within a generation, it saved billions of dollars by preventing floods. It helped farmers conserve productive soils. It spurred improvements in health, literacy, industrial production. It brought electricity, refrigeration, navigable links to seaports, and revenues from hydropower to devote to community development. It brought opportunity.

TVA's history underscores that investments in large-scale, geographically-based initiatives can pay off handsomely. They do require vision and leadership. Other elements of success: they are multiple use in purpose, they draw in the range of stakeholders, including the people who live there and are thus most affected, and they enjoy the active support of political leaders, national and local. Countries around the world at a stage of development comparable to the Tennessee Valley in the 1930s might benefit from this experience, including lessons learned about the need to consider environmental factors. And so might regional initiatives, in the Nile, Niger, and Senegal river basins, for example, supported by the Global Environment Facility and other international institutions.

Several of our colleagues at the meeting spoke directly about the need for water decisions to take account of the full spectrum of supply and demand issues in providing drinking water and sanitation. This notion of full accounting is embodied in the concept of integrated water resources management, which engages all the relevant functions, information, and stakeholders. Under this umbrella, good decisions can be made, or at least better decisions than in the past. Oftentimes, for example, conservation or improved efficiency of use is cheaper than developing new reservoirs or other expensive infrastructure. In fact, industry, power production, and agriculture are the largest consumers of water in the United States, and it is the improvement in efficiency in these sectors that underlies the US record of holding water use essentially level since the mid 1970s, even as population and economic production have soared.

Integrated water resources management can help ask and answer the right questions about meeting demand for water.

Internationally, a good example of the failure to apply integrated thinking can be seen in India. Farmers pay virtually nothing for electricity and thus in some communities are pumping so much groundwater that the water table is dropping 10 to 15 feet a year. That's hardly sustainable, and is especially lamentable when alternatives like drip irrigation are available to cut water demand and increase crop yields. Energy, agriculture, water - all would do better if the interrelationships were fully considered.

Though the economic and environmental cases for furthering adoption of integrated water resource management are becoming unassailable, many poor countries, we heard, need assistance from donor countries and agencies to define the problem fully (typically not merely increasing water supplies), and to set up the framework, institutions, plans, and financing to improve water resource management.

Public or Private?

Getting safe, affordable, and sustainable water and sanitation to those without are hardly contentious goals. But how that gets done can prompt considerable controversy. Two issues, in particular, seem to raise red flags - the idea that access to water should be a fundamental human right, even provided free to poor households, and the idea that the private sector can play a beneficial role in delivering needed services.

Arguments over privatization, in our view, distract from the fundamental objective, which is to improve water delivery and sanitation. This is a responsibility that belongs to governments and that many governments have manifestly failed to carry out. In the current environment, when long-term concessions to private companies are out of favor, and private investors are shunning water investments, we believe the focus must be directed at holding governments more strictly accountable for water services.

Some groups argue that access to water should be a fundamental right. We heard great sympathy around the table for ensuring access of all to clean water and sanitation. But the practical, operational dimensions gave everyone pause: What would it really mean? Who would enforce it? How?

That "water should be free" to some users also seems a popular refrain of some groups because of concern for poor households. Surely it's clear, however, that treating, transporting, and storing water costs money. So does treating and disposing of wastewater. Moreover, proper pricing encourages efficient use and reduces waste. Costs have to be covered in some way, by ratepayers through tariffs or by governments through tax revenues or a combination, and should be decided through a process that is transparent, accountable, and participatory and that ensures the needs of the poorest households are met. To be sure, full cost recovery may be an unattainable objective; few utilities, including in the United States, achieve it. Some may cover operations and maintenance, but few can cover capital costs for upgraded or expanded coverage. In fact, the US government helped finance the capital costs of water infrastructure throughout the country. So too, developing countries will have to look beyond ratepayers to provide funds for infrastructure.

The poor now pay dearly for their water - in payments to truck vendors, in ill health from drinking contaminated water, or in time spent securing water from distant sources. They also pay in the degradation of natural resources on which they depend. Concern for the poor is not misplaced. But practical and responsible methods for dealing with the problem are available - cheap rates for the amount of water necessary to sustain life, for instance, or transparent subsidies to poor households or water providers on their behalf.

A decade ago, private provision of water utilities was hailed by many as bringing new capital, management skills, and efficient operations to the global water challenge. But the approach hasn't realized its initial promise, and most private operators no longer see developing countries as good markets. The risks are large and rising, the returns rather limited. One major lesson from these experiences: the private model works only when there is a functioning legal and regulatory framework. This underscores the centrality of the governance agenda - transparency, accountability, anti-corruption, citizen participation, a working judiciary - to advancing better water services.

We embrace the notion that a quantity of clean water for drinking, bathing, and hygiene is necessary for life and for health, and providing it is a responsibility of government. That governmental bodies have failed to provide these essential water services, and to invest in their development and maintenance, led to private initiatives - and to a backlash. Opposition to private participation often obscures and confuses the essential challenges of providing water and sanitation: to create honest, transparent public bodies to oversee, or provide directly, water services, for which, in turn, they collect enough revenues and apply those funds to maintenance to keep pace with population growth, economic development, and other public needs. The typical experience of water departments in developing countries is one of overstaffing, poor accounting and billing, failure to maintain pumps and pipes, and recovery of only a third of the cost of service. This must be transformed.

Those who oppose private finance, construction, or management of water infrastructure have the obligation to help improve the delivery of public services or to explain how public bodies alone can do what most in the developing world have so conspicuously failed to, that is, provide adequate, clean water at reasonable cost. The urgent requirement is clean water. The question of who provides it, whether a public or private entity, strikes us as a secondary consideration.

It is often concern for the poor that causes groups to challenge private sector involvement in water, with mistrust of the profit-making motive perhaps underlying the concern. The controversy surrounding private water operators delivering services via ownership or contractual arrangements should not undermine the emergence of a new phenomenon: for-profit companies are joining with non-profit organizations in partnerships that voluntarily take on the challenge of helping provide water services to communities or institutions. More such companies are coming to understand that there is a strong direct business case for greater attention to water supply and demand, including the availability of clean water and sanitation to those in need. And more non-profit organizations, eager to use partnerships to achieve their aim, are coming to accept that businesses can make constructive contributions beyond philanthropy by developing and supporting new small-scale, sustainable, for-profit local enterprises in the water sector.

The company representatives who participated in the Wye dialogue recognize that without adequate supplies of clean water, they may not be able to continue operations. Without clean water, company products may not be usable. Without clean water, workers' ill health may lower productivity. Company representatives told us quite clearly that these practical incentives are causing more firms to consider how they might contribute to resolving water issues, whether leading initiatives that draw on company, community, and donor assistance; extending water services directly; contributing to nongovernmental or local groups to enable them to deliver services;

developing models for small-scale local enterprises to deliver water services; or providing other technical or financial help.

As more groups, institutions and companies - alone and in partnerships - address the need for clean water and sanitation for those who lack these services, we hope, and we expect, to see progress accelerate.

Recommendations

The discussions at Wye culminated in a series of recommendations that drew widespread support. (These are summarized at the beginning of this report.) Our intended audiences are the policymakers in Congress and the Executive Branch, who can make things happen directly or in concert with other governments and international agencies, and the broader community of interests around the US and abroad that have something tangible to contribute - non-governmental groups, companies, philanthropies, professional societies, and others. Many of these actors can move more quickly than governments or international agencies, and so we urge their direct involvement in meeting the global water challenge.

1) **Clean water and sanitation must become a higher priority because they are fundamental to human health and reducing poverty.** National governments, which bear prime responsibility, as well as regional and local governments, donors, and others in the water sector must provide greater resources and convey a sense of urgency. To monitor progress in meeting internationally agreed water and sanitation goals, periodic country-level reporting is needed, which will require assistance in countries without the ability to gather health-related statistics.

2) **All schools and orphanages should have clean water, sanitation, and hygiene education by 2015.** The United States and other donor countries, international agencies, developing country leaders and the business and non-profit sectors should mobilize resources to meet this need.

3) **The President of the United States and his Administration should develop a strategy to mobilize American resources and institutions to become more involved in water internationally.** The rationale for greater US involvement in meeting the need globally for safe, affordable, and sustainable water is compelling and is captured in legislation introduced by Senate Majority Leader Frist and co-sponsored by Minority Leader Reid and Foreign Relations Committee Chairman Lugar.

4) **For reasons of health, the economy, and environmental sustainability, governments must invest more in water infrastructure.** These investments must be considered in the context of other water related issues, including agriculture, energy, flood control, and ecosystem functions.

5) **Decisions about covering the costs of clean water and sanitation should be decided through a participatory process that ensures the needs of the poor are met and provides sufficient funds for maintenance.** Except for the poorest countries, the needed resources, which are substantial, for the most part will have to come from the affected countries themselves. Whether they are paid for by governments with tax revenues, by ratepayers through tariffs, or a combination, should be a pragmatic decision arrived at through a participatory process that is open, transparent, and accountable.

6) **Because water and sanitation are the responsibility of women in much of the developing world, they should become more directly involved in managing water resources and making water-related decisions.** Women currently bear most responsibility for collecting water for families in underserved communities and, along with children, will benefit most from better water and sanitation service. All agencies and institutions in the water sector should strive to ensure that women participate fully in managing and making decisions regarding water resources.

7) **Development assistance should emphasize building local capacity, creating legal frameworks for managing water, and building local sources of funding.** Improved municipal financial management in the developing world can enhance credit and expand access to domestic capital. Clear legal and regulatory regimes are essential for managing water and enabling private investment. Technical assistance is also necessary to build capacity for delivering water services. Rebuilding after natural disasters, when assistance may be more plentiful, should support sustainable solutions for water and sanitation services.

8) **Promising partnerships among governments, not-for-profits, community and faith-based organizations, and businesses should be replicated and scaled up.** Many creative interim and long term solutions to the need for clean water and sanitation exist. Mobilizing resources for these initiatives and coordinating their efforts are essential.

9) **Decentralized water treatment systems or point-of-use household treatment, coupled with sustained hygiene education, should be deployed more widely, especially where they can reduce water-related disease immediately.** These, along with

market-based, small-scale enterprises and other decentralized distribution and treatment options, offer promising new approaches to meeting the need for clean water and sanitation.

10) **Decisions about managing water resources must involve all stake holders and all relevant factors in supply and demand, with efficient water use and protection of ecosystems as central goals.** Planning efforts must take account of all aspects of supply and demand, including agriculture, energy, flood control, and ecosystem functions, as well as the needs of all interests.

The 4th World Water Forum

Besides seeking to stimulate greater US engagement on water issues internationally, the Aspen Institute and the Nicholas Institute convened this dialogue to contribute suggestions to the 4th World Water Forum scheduled for Mexico City, March 16 to 22, 2006. In January 2005, César Herrera Toledo, Vice Director of Mexico's National Water Commission General Program wrote, "We are pleased that the Aspen Institute's dialogue has been designed to serve as a Preparatory Workshop and look forward to receiving its results as we finalize plans for the Forum." The Commission has the lead in organizing the Forum for the government of Mexico.

As a contribution to the Forum, the following letter was sent April 20, 2005, to Vice Director Herrera.

April 20, 2005

Ing. César Herrera Toledo
Vice Director
National Water Commission / Comisión Nacional del Agua
Insurgentes Sur 2416, Col. Copilco el Bajo, C.P. 04340
Delegación Coyoacán, México D.F.

Dear Mr. Herrera:

As co-chairs of the "Water, Development, and U.S. Policy" dialogue convened by the Aspen Institute and the Nicholas Institute, March 30 - April 2, we are writing to report to you and other organizers of the 4th World Water Forum about the results of our meeting. Your January 6th letter to John A. Riggs at Aspen invited our contribution, which we are pleased to provide.

We had a full and stimulating discussion among some 30 participants, who came from a variety of sectors and backgrounds - environment, development, the public sector, private companies and more (a list of participants is attached). We will send the report and the website link when it is final. In addition to brief descriptions of nongovernmental, corporate, and other initiatives by U.S. groups working internationally to bring water and sanitation to communities, schools, and other venues, we anticipate the report will contain the following highlights:

- The need is clear to elevate the priority for clean water, sanitation, and hygiene education by governments, donor agencies, companies, and others; and progress is as much a matter of building political will as it is of building capacity and securing finance.

- As a signature initiative, agencies and institutions, including the private sector, should be mobilized to get water and sanitation to all schools and orphanages within a generation.
- The U.S. government should prepare a strategy for greater international leadership and participate constructively in the 4th World Water Forum.
- A large need exists for public investments to improve water resources management.
- Improving governance by creating a legal and regulatory framework for water resources management is as important as, and must accompany, greater funding.
- Except in the poorest countries, domestic sources of investment will have to cover most of the costs of water service improvements.
- Point-of-use and other community and household level interventions can offer immediate health benefits.
- Integrated water resources management can make an essential contribution to clean water and health, as well as other goals.

We are especially pleased that your colleague Francisco Gurria was able to join us to speak about preparations for the Forum. His presentation and remarks prompted an excellent discussion of the opportunities and concerns, which we summarize below.

First, with Mexico hosting this international Forum, a tremendous opportunity exists to shine the spotlight on the urgent need for clean water, sanitation, and hygiene education in Mexico, as well as throughout the developing world. Elevating the priority at home and among your neighbors in the region should be possible. In his opening remarks, President Fox has a unique platform to command world attention to this problem, speaking with passion and from a position of strength given all that you are doing in Mexico to address water issues. It is our understanding, for example, that Mexico is moving away from the notion of water as a free good, recognizing that clean water and sanitation have costs and those costs must be covered. Further, we understand that the new water law assigns responsibility for water management to state and local governments, and that the shift in irrigation from government control to user groups is nearly complete. These are important and timely moves and we urge that President Fox explain what Mexico is doing and why for the benefit of many of the assembled countries.

Second, the Forum's emphasis on solutions, tangible on-the-ground results, models, examples, case studies, and partnerships strike us as right on target, as embodied in the overarching theme "Local Actions for a Global Challenge." Past Forums have witnessed fundamental disagreements over philosophy, ideology, and strategy that have impeded progress, in our view. Though some of these issues may resurface, we hope they won't again distract from the most fundamental questions of how to improve access to clean water and sanitation to people in need and how to improve management of water resources for the benefit of all. Showcasing what works, why, and the lessons learned, we believe, will inspire others and make the point these water problems are solvable with sufficient will, finances, and know-how.

Third, communications and public relations planning are critical to the Forum's success. Not only can extensive publicity about the Forum advance the cause of clean water and sanitation in Mexico and worldwide, but we understand there may be an alternative civil society forum, which will be competing for media attention with the main Forum. We applaud your efforts to involve NGO leaders in the main Forum, at the Fair, on panels and in workshops, and in other venues. They often bring a rich experience of working directly with communities and local people, they bring a perspective that differs from that of government officials, and their involvement would be valuable. We encourage you to continue to seek their input in the planning and their participation during all or part of the Forum. One way might be to include a specific session or two on the key issues they cite - for example, water as a human right or private sector participation - to give them a voice at the Forum and structure the discussion.

News media, as no doubt you know, thrive on controversy and will try to exploit the differences between the main Forum and a civil society forum. There is no way to rein in top reporters, nor should you try. But providing them a steady diet of good experts and practitioners with whom to speak, project briefings, field visits, regular announcements, and story ideas and themes or messages for the day should help - in other words, a very active program for media. Remarks by President Fox and other prominent speakers throughout the Forum should help frame and underscore the issues. Former heads of state like Oscar Arias have enormous credibility and should attract attention. Other high profile celebrities or officials, likewise, can help present and amplify the Forum's messages.

Mr. Gurria stated that Forum organizers are working to find a few simple messages to come out of the Forum. This is welcome news. Our meeting recognized the need for a simple compelling message that conveys both the urgency of the water challenge and the ability to solve it, as a means of raising awareness and building public support. We were a technical and policy group, however, and did not arrive at any consensus about that message. We concluded that involving communications or public relations professionals would be worthwhile.

Fourth, Mr. Gurria asked us to suggest some U.S. examples that might be relevant. One is the Tennessee Valley Authority. In the 1930s, more than 90 percent of the people in this large, multi-state region had no electricity, about three-fourths had no piped water, few had radios, and less than a quarter of the people owned cars or trucks. Most lived on subsistence farming. Scil erosion and flooding were ruinous. Within a generation, that all changed. TVA saved billions of dollars by preventing floods. It helped farmers conserve productive soils. It led to improvements in health, literacy, and industrial production. It brought electricity, refrigeration, navigable links to seaports, and revenues from hydropower to devote to community development. TVA's history underscores that large public infrastructure investments can pay off handsomely, indeed they may be essential to spur economic growth. Countries around the world at a stage of development comparable to where the Tennessee Valley was in the 1930s might benefit from understanding this experience.

The United States also has considerable experience, including both successes and failures, that might usefully be shared about river basin management, large scale restoration and the importance of incorporating ecological values into water resources management. Our country has learned, often at extensive cost, that healthy, functioning ecological systems are critical to human and economic well-being.

Our recent experience with more efficient use of water may also be relevant to Mexico and Forum participants. In large measure due to improved efficiency of use in agriculture and industry, U.S. water consumption has remained in total about where it was in the mid 1970s and on a per capita basis about where it was in the 1950s.

Fifth, our personal experience on U.S. delegations to large international meetings suggests it may be too much to ask Ministers to forego negotiating an accord. Mr. Gurria stated that the Forum's hosts hope to draw a sizable official U.S. delegation, as well as many other participants from our country. We, too, hope that comes to pass. To help make it more likely, regarding the Ministerial Declaration, we urge that you use your government's good offices to ensure that the Ministers are not asked to negotiate new international rights or norms in the water sector. To ask for such negotiations, in our view, may well prove divisive and discourage high-level U.S. government participation.

There is much positive to say in a Declaration: Urging prompt action, reasserting internationally agreed to goals, recognizing the importance of partnerships, applauding commitments that have been made, underscoring the value of more efficient water use and the centrality of good water resource management to the health of people, economies, and the environment. These and other elements would seem fine and non-controversial. Too often, the Declaration is negotiated as if it is the most important output. Frankly, our hope would be that Ministers would not have to devote a majority of their time at the Forum to negotiating a Declaration on which there is disagreement over lengthy, bracketed text. It is the examples and models which will motivate others that most deserve time and attention.

We would hope to see the Ministers visit field projects, take part in workshops to learn what has worked, stay accessible to press, and meet with their counterparts and others to develop and launch initiatives to further extend water services to the poor and promote other Forum objectives. The partnerships that could emerge, given the opportunity for willing Ministers and others to meet, would be a far more exciting result from the 4th World Water Forum than another Declaration on water. In the end, our discussions led us to conclude that the time for talking about water problems has passed, the time for emphasizing solutions and action is now.

This 4th World Water Forum is an immense undertaking and all the dialogue participants join in wishing you and your fellow organizers great success. We would be pleased to discuss any of the above points at greater length if that is helpful.

We look forward to seeing you in Mexico City a little less than a year from now. With every good wish,

Sincerely yours,

William K. Reilly   Harriet C. Babbitt

Cc: Francisco Gurria
        Rodolfo Ogarrio

Participants

Co-Chairs:

Harriet C. Babbitt
Senior Vice President
Hunt Alternatives Fund
Washington DC

William K. Reilly
Founding Partner
Aqua International Partners
San Francisco CA

Participants:

Tralance O. Addy
President and CEO
Plebys International LLC.
Lake Forest CA

Gordon Binder
Aqua International Partners
c/o World Wildlife Fund
Washington DC

George D. Carpenter
Director, Corporate Sustainable
 Development
The Procter & Gamble Company
Cincinnati OH

Mike Connor
Minority Counsel
Senate Energy and Natural
 Resources Committee
Washington DC

Jerry Delli Priscoli
Senior Advisor for Water
 Resources
U.S. Army Corps of Engineers
Arlington VA

David Douglas
President
WATERLINES
Santa Fe NM

Alfred M. Duda
Senior Advisor, International Water
Global Environment Facility (GEF)
Washington DC

Sylvia Earle
Oakland CA

Monica Ellis
President and COO
Global Environment & Technology Foundation
Arlington VA

Gerry Galloway
Glenn L. Martin Institute Professor of Engineering,
University of Maryland
College Park MD

David Graham
Vice President
Environmental Health & Safety
Dow Chemical
Midland MI

Francisco Gurria
Consejo Consultivo del Agua, A.C.
México D.F.

Henry Habicht
CEO
Global Environment & Technology Foundation
Arlington VA

G. William Hoagland
Director of Budget and Appropriations
Office of the Senate Majority Leader
Washington DC

Karin Krchnak
Director of International Water Policy
The Nature Conservancy
Arlington VA

Peter Lochery
Senior Water Advisor
CARE USA
Atlanta GA

Kathryn Oakley
Head of Public Affairs
RWE Thames Water
London UK

Tim Profeta
Director
Nicholas Institute for Environmental Policy Solutions
Duke University
Durham NC

Ken Reckhow
Professor, Nicholas School of the Environment & Earth Sciences
Chair, Environmental Sciences & Policy
Duke University
Durham NC

Peter Reiling
Vice President for International and Policy Programs
The Aspen Institute
Washington DC

John A. Riggs
Executive Director
Program on Energy, the
  Environment and
  the Economy
The Aspen Institute
Washington DC

Aaron A. Salzberg
Senior Advisor on Water
Office of Policy Coordination and
  Initiatives
US Department of State
Washington DC

Jacqueline E. Schafer
Deputy Assistant Administrator
Bureau for Economic Growth,
  Agriculture and Trade
US Agency for International
  Development
Washington DC

Jeff Seabright
Vice President, Environment and
  Water Resources
The Coca-Cola Company
Atlanta GA

Thomas G. Searle
President
Water Business Group
CH2M-Hill
Englewood CO

Frank Tugwell
President and CEO
Winrock International
Arlington VA

Mark Van Putten
Principal
ConservationStrategy™
Reston VA

Daniel Vermeer
Director, Global Water Initiative
The Coca-Cola Company
Atlanta GA

Steve Werner
Executive Director
Water for People
Denver CO

Jerry Wiles
President
Living Water International
Houston TX

Evening Speakers:

Holly Burkhalter
US Policy Director
Physicians for Human Rights
Washington DC

John F. Turner
Assistant Secretary
Bureau of Oceans and
  International Environmental and
  Scientific Affairs
Department of State
Washington DC

For Further Information

A reader concerned about the silent tsunami might well ask: what can I do? The participants at Wye offered a few thoughts.

First, let your Senators and Members of Congress know you care about US involvement in water issues internationally and ask what they are doing to help.

Second, consider supporting community-based, faith-based, and other groups that are delivering water services to people in need. Though hardly an exhaustive list, the groups represented at Wye offer a starting point:

- CARE, www.careusa.org/careswork/whatwedo/health/water.asp
- Living Water International, www.water.cc
- Millennium Water Alliance, www.mwawater.org/key.html
- Water Advocates, www.wateradvocates.org
- Water for People, www.waterforpeople.org
- WaterHealth International, www.waterhealth.com
- Winrock International, www.winrock.org/what/forestry.cfm

Third, learn more about water issues. Again, though hardly exhaustive, the following websites are considered informative:

- UNESCO, www.unesco.org/water/
- UNICEF, www.unicef.org/wes/index.html
- US Environmental Protection Agency, www.epa.gov/ebtpages/ water.html
- The World Bank, www.worldbank.org/watsan/
- Pacific Institute for Studies in Development, Environment, and Security, www.pacinst.org/topics/water_and_sustainability/
- The World Water Council, www.worldwatercouncil.org
- The 4th World Water Forum, www.worldwaterforum4.org.mx